W9-CGY-352

SCIENTIFIC KNOWLEDGE AND PHILOSOPHIC THOUGHT

SCIENTIFIC KNOWLEDGE & PHILOSOPHIC THOUGHT

Harold Himsworth

THE JOHNS HOPKINS UNIVERSITY PRESS
Baltimore and London

The Johns Hopkins University Press,
701 West 40th Street, Baltimore, Maryland 21211
The Johns Hopkins Press Ltd., London

The paper used in this publication meets the minimum
requirements of American National Standard for Information
Sciences—Permanence of Paper for Printed Library
Materials, ANSI Z39.48-1984.

Library of Congress Cataloging-in-Publication Data

Himsworth, Harold, 1905–
    Scientific knowledge and philosophic thought.

    Bibliography: p.
    Includes index.
    1. Science—Philosophy.   2. Science—Methodology.
3. Knowledge, Theory of.   4. Problem solving.   I. Title.
Q175.H48   1986   501   85-24118
ISBN 0-8018-3316-7 (alk. paper)

Contents

vi

Contents

Foreword

Harold Himsworth, as a scientist and as a physician, sees reality through facts and their logical consequences. In this small and elegant book, he focuses on how philosophers use words and their connotations for establishing truths. Believing that we are more likely to reflect reality by seeing philosophic problems in straightforward ways, he rejects as misguided those arguments that reach for profundity through reasoning based upon unrealistic premises. In particular, Himsworth has no faith in attempts to discern how we think by mere use of words as opposed to careful observation.

In the future, most likely far distant, we shall perceive logic as a branch not of philosophy but of neurobiology, in much the same way that, beginning with Galileo and Newton, natural philosophy became transformed into physics. Until then, the scientific world is bound to be intrigued by philosophers concerned with how ideas are generated. But here we must not automatically assume that, because their arguments are increasingly subtle, they represent serious advances beyond the commonplace ideas about deduction and induction first formulated so well some three hundred fifty years ago by Francis Bacon. I, like Harold Himsworth, am uncomfortable with much of this unneeded complexity. To see its pitfalls, as well as to enjoy the urban civility of Himsworth's thought, I most enthusiastically endorse the reading of the words that follow below.

J. D. WATSON

Cold Spring Harbor Laboratory
December 1985

SCIENTIFIC KNOWLEDGE AND PHILOSOPHIC THOUGHT

ONE
Methods of
Thought

Human knowledge is ordinarily classified according to the matters with which it deals. On this basis, three broad categories are distinguished: the scientific, the theological, and the philosophic. The first deals with natural phenomena, the second with supernatural. It is less easy, however, to specify the concerns of the third. Confronted by this difficulty at the outset of his *History of Western Philosophy,* Bertrand Russell writes: "Philosophy, as I shall understand the word, is something intermediate between theology and science. Like theology it consists of speculations on matters as to which definite knowledge has so far been unascertainable: but like science it appeals to the human reason rather than authority, whether that of tradition or that of revelation." And, he goes on, "All *definite* knowledge—

or so I should contend—belongs to science; all *dogma* as to what surpasses definite knowledge to theology. But between theology and science is a No Man's Land exposed to attack from both sides: this No Man's Land is philosophy."[1]

But what are we to understand by these statements? On the face of it, Russell seems to be suggesting that the distinguishing feature of philosophy is one of method rather than substance. If this be so then we are left with two possibilities. Either there are three kinds of problem—the scientific, the theological, and the philosophical—each requiring a different method for its solution; or there are three different ways of solving a problem, and according to which we adopt we shall get a different answer.

If there are three kinds of problem, there is nothing more to be said. We must accept that this is the position and attribute failure to make progress in any of these fields to the greater difficulty of its problems. If, however, there are three different ways of solving a problem, the position is different. Now we must contemplate the possibility that failure to advance knowledge in a particular field may be due, not to its problems being more difficult, but to inadequacies in the method we customarily use for solving them. In view of the relative rates of progress in the three traditional fields of knowledge over the last two thousand years, this second possibility is not one that can be dismissed out of hand. Accordingly it would seem important to look into it further.

Scientific Knowledge and Philosophic Thought

Progress in knowledge depends upon the development of ideas. Starting with simple ideas—which are no more than labels for the impressions made on the mind by items of experience—the imagination proceeds, by combining them, to arrive at concepts of more complex objects or events. Then, by combining these in turn, it constructs still more complex ones. For instance, if I have the idea of gold and the concept of a crown, I can conceive of a golden crown. If I have the concept of a man and the concept of death, I can conceive of a human corpse. If I have the concept of a house and that of a ghost, I can conceive of a haunted house. All this I can do without ever having come across any of these things. For, as David Hume said, "nothing is more free than the imagination of man and, although it cannot exceed the original stock of ideas furnished by the internal and external senses, it has unlimited powers of mixing, compounding, separating, and dividing ideas in all varieties of fiction and vision."[2]

To the process of concept formation, therefore, there is no end. Moreover, being a purely intellectual operation, it can proceed without recourse to further experience. Yet there is one proviso, and this is absolute. If a concept is to be credible, the ideas or simple concepts out of which it is constructed must not contradict one another. Thus, I cannot conceive of an object being in two different places at the same time. Logically speaking, such a concept would be nonsensical, and, as such, the mind rejects it out of hand.

In this respect, the method of thought of the theologian, the scientist, and the philosopher are at one. All make use of logic to process the ideas or concepts from which they start. It is not at this stage, therefore, that we should look for any differences between them. Rather it is at that which precedes this.

Broadly speaking, the initial concepts from which logical thought takes its start are of three kinds: postulates, statements of fact, and propositions. These correspond respectively to the raw materials of theology, science, and philosophy.

A postulate is a concept, acceptance of which is mandatory. As such, it determines all subsequent thought in its field. All experience is interpreted in the light of its precepts. All other concepts are accepted or rejected according to whether they are compatible with it or not.

In contrast are concepts deriving from factual observation. These differ from postulates in that they depend for their acceptance, not on authority, but on being demonstrably in accord with fact. Such are those that lie at the base of the body of knowledge we call scientific.

Propositions lie between these two extremes. They differ from postulates in that the individual is not required to accept them without question; from scientific statements in that the concepts they embody have not been subjected to a comparably rigorous process of factual verification. Their title for

Scientific Knowledge and Philosophic Thought

acceptance rests upon their being logically irrefutable. They are, so to speak, the end of the road where logic is concerned, in the sense that they cannot be reduced by logical analysis to anything simpler.

Looking at these three categories of concept, it will be seen that those in the first differ essentially from those in the other two. Being postulates, their validity must be taken on trust. Factual statements and propositions, on the other hand, can only be accepted if they satisfy certain requirements. In short, the concepts that lie at the basis of thought in science and philosophy are subject to human judgment, whereas those that do so in theology are not. In inquiring into human knowledge, as distinct from belief, it would seem proper, therefore, to exclude the latter from consideration.

But that leaves us with the problem of the status of philosophy as a branch of knowledge. Does it follow that because a proposition is logically irrefutable it is necessarily valid? Might not the concept it embodies be at fault, either because it fails to take account of all the relevant information or because it has been insufficiently tested against experience? If so, although the proposition in question may well serve as a basis on which to construct a logically impeccable system of thought, this may bear little relation to reality. It would seem worthwhile, therefore, to look at some of the propositions from which philosophy takes its start from this point of view. This we can best do by first considering one that lies close to the frontier between philosophy and science.

Methods of Thought

TWO

Experience & Understanding

In the middle of the eighteenth century, David Hume produced a disquisition on human understanding that has disconcerted philosophers and laymen alike ever since. And no wonder, for it cut at the root of the belief on which all our knowledge rests; namely, that in the light of past experience, it is possible to infer from present events what will happen in the future.

On the face of it Hume's contention seems opposed to all common sense. And Hume was no more devoid of common sense than he was of intelligence. He confessed that he framed his own life on the assumption that the past is a guide to the future. Furthermore, he recognized that the whole of human knowledge rests on the belief that there is a causal relationship between events. All this he conceded with his customary urbanity. Yet, he was unable to fault his own logic, and as Bertrand Russell pointed out, subsequent philosophers have been no more successful.[1]

We are thus left in a curious position. On the one hand, we see human achievements in the scientific sphere increasingly justifying our belief that, on the basis of previous experience, we can reason from cause to effect and vice versa. On the other, we have as yet failed to refute Hume's contention that such a belief is logically untenable.

Is it not possible, however, that in seeking to refute Hume's arguments we have been misdirecting our efforts? Might it not be that what was at fault was not his logic but the proposition from which he started?

II

The argument underlying Hume's contention runs as follows: "For all inferences from experience suppose, as their foundation, that the future will resemble the past and similar powers will be conjoined with similar sensible qualities. If there be any suspicion that the course of nature may change and the past be no rule for the future, all experience becomes useless and can give rise to no inference or conclusion. It is impossible, therefore, that any argument from experience can prove this resemblance of the past to the future, since all arguments are founded on the supposition of that resemblance."[2]

Given the proviso in the second sentence of the foregoing statement, the conclusion in the third is logically irreproachable. But, if we can suppose that the course of nature might change, we can equally well suppose that it might not. In that case we should have to rewrite the second and omit the third sentences of Hume's paragraph so that it reads as follows: "If there be no suspicion that the course of nature may change and the past may not continue to be the rule for the future, therefore, inferences and conclusions can properly be drawn from past experience as to the course of future events."

Thus, according to whether we start from the proposition that the course of nature might change, or the proposition that it might not, logic will lead us inexorably to diametrically opposite conclusions. If we opt for the first of these propositions, we shall be driven, like Hume, to conclude that it is impossible to argue from past experience and that our belief in cause and effect is misplaced. If, on the other hand, we opt for the second, we shall be driven, equally firmly, to conclude that it is possible to reason in this way and that our belief in cause and effect is fully justified. Given the proposition from which the argument starts, both these conclusions are equally logical. It is impossible, therefore, to decide between them on this basis.

But does this rule out the possibility of deciding between them on other grounds? After all, if the course of nature were to change, that would be a fact. The question whether it has or has not done so is, consequently, one that falls to be decided on factual grounds. In order to take the measure of the proposition that it might change, therefore, should we not consider what would happen if it did?

Experience and Understanding

Clearly, the course of nature could not change unless the forces that had hitherto determined its course were themselves to do so. Consider, therefore, what would happen if one such force, gravity, were to stop.

If I toss a stone into the air, then, on the basis of previous experience, I expect it sooner or later to fall to the ground. If, however, the force of gravity were to fail, it would not. It would continue on into outer space. It would, in Hume's words, have lost its "sensible quality" of heaviness and, hence, its "power" of returning to earth.

This, however, gives a wholly inadequate picture of what would happen if gravity were to stop. The effects of its doing so would not be confined to any particular kind of object. Everything that has weight would be affected. For instance, this planet would no longer be able to retain its atmosphere. As a result, all living organisms that depend on having air to breathe would die, and there would be none left to experience anything. The fact that there are such creatures alive today means, therefore, that as long as they have existed, gravity has been operating. Furthermore, that as long as they continue to exist, it will not cease to do so.

But the force of gravity is not a thing apart. The same considerations apply to all other forces that govern the course of nature. If those that determine the course of chemical reactions, for instance, or the transformation of one form of energy into another, were to fail, all life on Earth would come to a halt. In short, the form and functioning of all living creatures have been dictated by the natural forces that exist and, should any of these cease to operate, all such creatures would, inevitably, die. Although, therefore, we cannot be sure that these forces will *never* fail, we can be certain that, if they do, nobody will be alive to witness the event. In consequence, the proposition that the course of nature might change, although logically irrefutable, is irrelevant to any inquiry into the development of human understanding.

The conclusion that follows from these considerations is clear. As long as human beings exist, they can count on the future resembling the past and, hence, can justifiably infer the course of future events from past experience.

Would it be wrong to say, therefore, that Hume set out to determine the limitations of human knowledge but what he actually succeeded in doing was to demonstrate the limitations of abstract thought, however logical, as an instrument for advancing this? In other words, the fault in his argument lay, not in his logic, but in his failure to take into account the full implications of the proposition to which he applied it.

Experience and Understanding

THREE

Observations &
Hypotheses

In view of the achievements of scientific research over the last three centuries, it may come as a surprise to learn that it is now being questioned whether the method scientists claim to use in advancing their knowledge is logically tenable. Accustomed as we are to regard logical justification as the ultimate test of rationality, this is disconcerting. It would seem important, therefore, to look at the objections now being raised.

II

The traditional view of the way scientific knowledge develops is identified with the name of Francis Bacon. Living as he did in the aftermath of scholasticism, Bacon was well aware that it is "in the nature of the mind of man (to the extreme prejudice of knowledge) to delight in the spacious liberty of generalities . . . and not in the inclosures of particularity."[1] So he set himself to lay down guidelines for advancing natural knowledge which proceeded, not from the general to the particular, but from the particular to the general.

That men could arrive at general ideas about things on the basis of observing many particular instances of these—that is, by thinking inductively—was commonly accepted. But, as Bacon appreciated, "to conclude upon an enumeration of particulars, without instance contradictory, is no conclusion, but a conjecture; for who can assure (in many subjects) upon those particulars which appear of a side that there are not others on the contrary side that appear not."[2] So, he tried to devise a better kind of induction than what is called induction by simple enumeration. The first step, as he saw it, was to construct "a Kalendar, resembling an inventory of the state of man, containing all the inventions (being the works or fruits of nature or art) which are now extant, and whereof man is already possessed; out of which naturally result a note, what things are held impossible . . . which Kalendar will be the more artificial and serviceable, if to every reputed impossibility you add what thing is extant which cometh nearest in degree to that impossibility; to the end that . . . man's enquiry be more awake in deducing direction of works from the speculation of causes."[3] Thus, for example, he hoped that by considering many hot bodies and many cold ones, the observer would be led to identify some common

feature of those that were hot and so arrive at a
general concept of the nature of heat.[4]

This, however, Bacon regarded as only the first, or "preparatory," stage in the process of "invention" (i.e., arriving at a hypothesis about something). The second, which he called "suggestion," is to test the hypothesis thus formulated. "Suggestion," he writes, "doth assign and direct us to certain marks, or places, which may excite our mind to return and produce such knowledge as it hath formerly collected, to the end that we may make use thereof. . . . But the same places, which help us what to produce of that we know already, will also help us . . . what questions to ask."[5] "For as in going a way we do not only gain that part of the way which is passed, but we gain the better sight of that which remaineth; so every degree of proceeding in a science giveth a light to that which followeth."[6]

Translating Bacon's phraseology into modern form, the traditional view of how scientific knowledge advances is broadly as follows.

From his observations of things and happenings in the world around him, man* acquires a knowledge of individual facts. By bringing together the presently available facts that bear on a particular problem he may—hopefully—see their general im-

---

*Throughout this book the terms *man* and *men* are used in their generic sense. Neither they, nor their corresponding pronouns, should be taken, therefore, to refer exclusively to a male member or members of the human race unless the context indicates that this is so.

plications and, on this basis, formulate a hypothesis regarding its explanation. He is then in a position to deduce what might logically follow, if his hypothesis were correct, and look to see if it does. Should his expectations be borne out by his further findings, his confidence in the validity of his hypothesis is correspondingly fortified. If, however, they are not, he must either amend it to take account of the additional facts he has discovered or, if that is impossible, abandon it altogether.

The foregoing procedure involves, therefore, two successive steps: that in which hypotheses are formed; and that in which they are put to the test of further experience. In the former the individual proceeds from the particular to the general, that is, inductively; in the latter, from the general to the particular, that is, deductively.

From the standpoint of logic, the second of the above steps presents no problem. The hypothesis under test is simply a premise, and it is only an exercise in logic to deduce what would follow if it were correct. But it is otherwise in regard to the first step. As yet, it has proved impossible to produce a logical justification for inductive argument. In the opinion of certain philosophers, notably Karl Popper and his school, this constitutes an insuperable obstacle to accepting the traditional view of the way in which scientific knowledge is developed.[7] Accordingly, they have put forward another which seeks to avoid the difficulty by dispensing with any need to invoke the aid of induction. Instead, they

contend that hypotheses are essentially intuitions,[8]
and that the role of factual evidence is confined to
the second stage, that in which hypotheses are put to
the test. In Popper's words, scientific knowledge
advances by a process of "conjecture and refutation." In short, it is only in part a rational
procedure.

This is a startling suggestion. It is important,
therefore, to see how such a view has come to be
held.

III

The belief that the role of factual experience in the
advancement of scientific knowledge is confined to
that of refuting or corroborating a previous conjecture rests on the contention that observation is
never fortuitous. It is always directed. "Observation," Popper writes, "is always selective." "The
belief that we can start with pure observation alone,
without anything in the nature of a theory, is absurd."[9]

Once this proposition is accepted, it follows
automatically that hypotheses are the products, not
of reason, but of intuition. The argument is as
follows.

Believing as he does, Popper is immediately
brought up against the problem of "Which comes
first, the hypothesis (H) or the observation (O)?"
This he says is soluble in the same way as the

Observations and Hypotheses

problem, "Which comes first, the hen (H) or the egg (O)?" The reply to the latter, he writes, "is an earlier kind of egg; to the former, an earlier kind of hypothesis."[10] On this basis, if asked what comes before that, he perforce has to answer, a still earlier hypothesis. And so ad infinitum.

This line of argument, however, leads nowhere. At no stage are we any nearer answering the question we set out to answer. We have become firmly entangled in what, in logic, is known as an infinite regress. From this Popper endeavors to escape by postulating the existence in man of certain "unconscious *inborn* expectations." He rejects the "theory of inborn ideas as absurd." "But," he writes, "every organism has inborn *reactions* and *responses;* among them responses to impending events. These responses we may describe as expectations...." "Thus we are born with expectations; with 'knowledge' which, although not valid *a priori,* is psychologically or genetically *a priori,* i.e. prior to all observational experience. One of the most important of these expectations is the expectation of finding a regularity. It is connected with an inborn propensity to look out for regularities, or with a *need* to *find* regularities."[11]

But is there not a far simpler explanation, and one that removes any necessity to postulate any such thing?

Suppose, for a moment, the proposition that observation is always directed by a pre-existing hypothesis is mistaken. Suppose that sometimes, at least, it is purely fortuitous. In that case, if asked which comes first, the hypothesis or the observation, we should have had to answer, "the observation." In tracing the development of any line of thought we should then have been driven back, step by step, until we reached a stage at which a fortuitous experience had forced an observation on man's attention which previously had been totally unsuspected, and about which, in consequence, he could have had no hypothesis.

This line of argument completely removes any necessity to postulate anything like an intuitive sense which, "prior to all observational experience," leads men to see regularities in natural phenomena. For was there any reason for postulating the existence of such a sense save the need to escape from the infinite regress into which one is betrayed by the view that "observation is always selective"?

It would seem, therefore, that here again we are faced with a situation in which logic leads us to radically different conclusions according to the proposition from which we start. And, in this case also, the question of which of the alternative propositions is correct is one of fact. As such, it is one that cannot be decided by logic but only in the light of the evidence.

IV

Undoubtedly, Popper is correct in saying that, once possessed of a hypothesis about anything, scientists tend to look preferentially in a particular direction for further information. But, as all experienced research workers know, there are other occasions (and these among the most fruitful), when a completely unforeseen observation is forced upon their attention and overturns all their preconceptions. In the history of scientific thought there have been many such instances.[12] A few examples will suffice to establish this point and, incidentally, to dispose of the suggestion that, although the observation in question might seem unpremeditated, the observer was subconsciously predisposed to make it. The first I inadvertently repeated on myself when a young man, the second happened to one of my predecessors, and the third occurred in a research institute with which I was then concerned.

After prolonged efforts, the Curies, man and wife, succeeded in isolating a small quantity of radium. So anxious were they to safeguard this that Pierre Curie carried it about with him in his waistcoat pocket. After a week or so he became aware that the skin of his abdomen under the pocket was itching and, on examination, saw that it was inflamed. Guessing the cause, he removed the radium. But the inflammation took time to subside and left him (as it has left me) with the typical scar of a radiation burn. That was the origin of our knowledge of the damaging effects of nuclear radiation on living cells and the start of the subject of radiobiology.

At the turn of the century, Sydney Ringer, a previous occupant of the chair of medicine I once held, was engaged in a series of experiments in which isolated frogs' hearts were perfused with saline solutions. Being a meticulously careful person, he insisted that only doubly distilled water should be used in making up these solutions. This duty he entrusted to his technician. Periodically, however, his technician became ill and had to go into hospital. In those penurious days, this meant that the professor had to make up the solutions for himself. To his dismay, Ringer then found that the perfused hearts quickly weakened and stopped beating. No sooner did his technician return, however, than the hearts beat as strongly and for as long as before. After this had happened on several occasions, Ringer's suspicions were aroused, and he cross-examined his technician closely. It then emerged that the latter had been saving himself trouble by making up the solutions with tap water, on the grounds that London tap water was good enough for anybody. So the influence of calcium on muscular contraction was revealed and the way opened to developing our present knowledge of the role of inorganic ions in the body's economy.

More recently a somewhat similar incident occurred in connection with the problem of preserving living tissues. If living cells from higher animals are frozen, they are killed. When, therefore, a claim was made that, if they were frozen in a solution of fructose they would be found on thawing to be alive, it aroused much interest. Unfortunately, other workers were unable to confirm this finding. One,

Observations and Hypotheses

however, after repeated failures, suddenly found that he could do so. Fortunately, before exhausting the contents of the bottle labeled "concentrated solution of fructose," he thought to have them analyzed. They were then found to be glycerol. How this situation came about was never established. Clearly, however, either glycerol had been put into the wrong bottle or, more probably, the labels on the bottles had become detached and some well-intentioned person had stuck them back onto the bottles to which he or she thought they belonged but, in doing so, had transposed the labels for fructose solution and glycerol.

Admittedly, such unequivocally fortuitous experiences are uncommon. But enough are on record to dispose of the belief that they never occur. And, as logicians never tire of pointing out, even one contrary experience is sufficient to refute a generalization that allows of no exceptions. To cite their favorite example, the statement that all swans are white becomes untenable the moment we see a black one. In the same way, the contention that observation is always directed by a previous hypothesis is disproved the moment we come across one that patently was not.

But consider what this means. In each of the above instances a completely unexpected observation set in train an entirely new line of thought and led to the development of hypotheses where none had existed before. Infrequent as such experiences are, therefore, they are of decisive importance for any understanding of how scientific knowledge develops, for they demonstrate unequivocally the seminal role of observation in the genesis of hypothesis formation.

The foregoing examples, however, are but extreme instances of a general principle. Take the observations we are led to make as a result of an invention that extends our range of vision.

One of the most significant observations ever made was that of Galileo when, consequent upon the invention of the telescope, he saw that the planet Jupiter possessed moons. This led to the formation of entirely new concepts about the structure of the universe. Yet even Galileo could have had no inkling of what he might see if he directed his telescope at that planet. All he knew was that, given an instrument with its potentialities, he might see something new.

Similar considerations apply to the legacies of further information that followed on the invention of the light microscope, the spectrometer, x-rays, and so on.

In my lifetime, I have seen such developments follow the introduction of the electron-microscope into the biological field. Recognizing the vastly increased range of observations this instrument brings within their grasp, research workers everywhere hastened to apply it to their materials. In doing so the great majority, at least, had no idea

Observations and Hypotheses

what it might reveal. All they knew, at the outset, was that it might turn up something that had a bearing on their problem. In the event, they were not disappointed. The result has been a great leap forward in our understanding of the structural basis of cellular processes.

These considerations are, however, in no way confined to observations that are accidental or incidental to technical development. They apply also to most of those men are led to make by virtue of holding a preconceived hypothesis.

As Medawar has pointed out, further observation seldom wholly discredits an existing hypothesis. Usually it does so only in part. All that is then required is to amend one's hypothesis to take account of the qualitatively or quantitatively unexpected element in the new finding.[13] Clearly, however, if the investigator had stumbled across this particular observation before formulating his hypothesis, this would have appeared, not in the form it did, but as it was subsequently amended.

We should recognize, therefore, that, in real life, observations range from those that are totally unexpected to those that accord entirely with expectations. The vast majority, however, fall in between. In other words, the exceptional observation is one that contains no unexpected, and consequently unseekable, element. Indeed, if this were not the case, there would be no point in doing research.

It appears, therefore, that the new view as to how scientific knowledge develops is based upon a factually mistaken premise. The traditional, on the other hand, is consistent with what actually occurs. Men do argue from the particular to the general, and, judging by the achievements of scientific research, do so successfully. The question is, on what basis do they do this?

Observations and Hypotheses

FOUR

The Particular
& the General

I f human beings are to have any understanding of their circumstances, they must reduce the multitude of their separate experiences to some kind of order. This they do, in the first place, by classifying objects according to their similarities. Thereby they arrive at a series of generalizations which, by defining the properties of objects of a particular type, enable them to recognize instances of these when they meet them again.

But the number of individual properties in nature is limited. Consequently, a variety of different objects may have one in common. Thus, there are many that are white, many that are solid, many heavy, many rounded, many warm, and so on. No particular object, however, possesses only one property. All have several. Ice, for example, is a transparent, colorless solid that is markedly cold to the touch. If held in the hand, it melts and turns to water. Glass also is solid and transparent and, unless tinted, colorless. But it is no colder than its surroundings, nor does it melt at body temperature. Again, apples and pears are solid objects and, if of the same size, about equally heavy. Each, however, has a different shape and taste. In none of these instances does any single property alone provide a certain identification of the object in question. For this, all are required. It seems, therefore, that the similarities with which men are concerned when classifying objects relate, not so much to single properties, but to particular combinations of these.

Now, the question as to whether an object possesses a particular combination of properties is as much a matter of fact as whether it possesses any one of them. As such it is one that can only be settled by observation. Similarly, the question whether several objects possess the same group of properties is one that can only be answered in the same way. In other words, observation supplies us with information, not only about the presence or absence of a particular property in a particular object but also about the particular combination of properties it possesses. All we then have to do is to put this into words in order to obtain an approximate idea of its distinguishing characteristics. We are then in a position to recognize such an object in the future and, on the basis of the extended experience thereby acquired, to arrive at a generalization about objects of this particular kind.

It would seem, therefore, that, at the level of the classification of objects at least, the process of generalization depends, not on thought, but on factual observation.

II

But observation of an object in isolation tells us no more than that it exists. Seen in relation to something else, however, it may be revealed to have certain implications. For example, if I apply a light to a pile of sticks, I expect this to catch fire and, shortly, to see smoke arising from it. If I see a rapidly traveling billiard ball make contact with a free-standing stationary one, I expect the latter to start moving. If I put a pan of water on the stove, I expect that, sooner or later, its contents will boil and turn into steam.

Why do I expect these things? Obviously, because previous experience (my own and that of others) has led me to do so. Without this there would be nothing to prevent my thinking that one could have smoke without a fire, that billiard balls would start moving of their own accord, or that water would boil in the absence of heat. As it is, observation has shown me that this is not so. For any of the above particular things to come about, it must have been preceded by another equally particular thing.

Thus, the order in which things succeed each other in nature is as much a matter of fact as the existence of the things themselves. So, just as in the

The Particular and the General

case of the combination of properties which characterizes a particular object, observation supplies men with information, ready-made, about what will happen if one kind of object comes into contact with another. Again, all they have to do is to put this into words in order to obtain an idea which will enable them to recognize such a sequence in future. Then, in the light of this further experience, they can formulate a generalization that applies to all sequences of this kind. Thus equipped they can now infer, on seeing the first item in a sequence, what would follow if it were to come in contact with the second. Equally, on seeing the second, what must have preceded it. In short, they can now generalize regarding the implications of the objects in question.

This is the basis for man's belief in cause and effect. David Hume put the matter succinctly when he said that, as a result of experience, man comes to believe that "like objects placed in like circumstances will always produce like effects."[1] Unfortunately, however, he had by then become committed to the view that, because we could not logically exclude the possibility that the course of nature might change, we were not justified in believing that an object seen today would produce the same effects as on a previous occasion.[2] If, however, the fact that men are alive to make observations shows that no such change has occurred, there is clearly no substance in this objection. Consequently, men are justified in thinking that any generalization they make regarding the implications of a particular thing will, insofar as it reflects factual observations, continue to be true as long as human beings exist.

III

It would seem, therefore, that, within the limits of our experience, we can formulate two kinds of generalization. The first, "Descriptive Generalizations," which necessarily precedes the other, relates to the combination of properties which characterizes a particular type of object; the second, which for reasons that appear in the next chapter we can call "Explanatory," concern the effects of different objects on each other. Both derive from factual observation, and thought only enters into their formation insofar as it is required to put into words what has been observed.

But if generalizations derive from factual experience, it follows that they can be no more valid, and no more comprehensive, than the observations on which they are based. Accordingly, it is of prime importance to man that his observations should be, not only free from error, but also complete. The question of the extent to which this is possible is, consequently, central to any inquiry into human understanding.

FIVE

Possibility &
Certainty

In the case of descriptive generalizations relating to the combinations of properties that characterize objects within our experience, there is no danger, human error apart, of their being factually incorrect. They are but verbal representations of what has been observed. Thus, ice *is* cold. It *is* colorless. It *is* transparent. It *is* solid. If warmed it *does* melt and turn to water. These are indisputable facts and permit of no equivocation.

But such generalizations may be at fault in another way. They may be correct in a limited context but incorrect in a larger. At one time, for instance, the generalization that swans are white birds of a particular shape and size that swim on water was perfectly acceptable to the inhabitants of Western Europe. When some of them discovered Australia, however, they saw birds with all the appearances of swans save that they were black. Consequently, they were driven to amend their previous generalization by eliminating color from the combination of attributes they had hitherto regarded as characteristic of this family of birds.

This example illustrates the way in which human knowledge of objects progresses from the particular to the general. It does so by eliminating from a generalization about a group of objects those properties that are not common to all until there remain only those that are. In effect, it proceeds by a succession of approximations, each one of which is dependent upon a further widening of experience. But no generalization about the pattern of properties which characterizes a particular category of objects can be universally valid unless it is based upon a complete experience of all the objects of this kind. Although, therefore, man can formulate descriptive generalizations that are valid within the limits of his experience, he cannot assert that they are universally so, for further experience may reveal them to be defective in some respect.

This raises the question as to whether we, by taking thought, can transcend the limits of our experience. Could the inhabitants of Western Europe have foreseen that, when they came to discover other parts of the world, they would see swans that were black? There was nothing to prevent them from fancying that they might. Obviously, however, they could have no certainty that such birds existed until an observation, incidental to their discovery of Australia, revealed that this was actually so.

We should frankly recognize, therefore, that the

degree of certainty attaching to a *descriptive* generalization depends upon the completeness of the factual observations on which it is founded. To be absolutely certain, it would need to be based on all past, present, and future objects of the kind in question. That would clearly be impossible. Nevertheless, we should not allow this to mislead us into discounting the validity of a generalization in its own context. The fact that some swans are black, although disproving that all are white, in no way detracts from the certainty that those in a particular subgroup are white.

This consideration is of importance when we come to the second category of generalizations— those I have called "Explanatory"—for here we are concerned with how some *particular* thing comes about.

II

The sole ground for thinking that two things are related is the observation that one tends to follow the other. It is accordingly important to know whether such an apparent association is real or fortuitous.

But can man ever be certain that a real connection exists between two different things, however often one follows the other? If not, what is the status of frequency of association in promoting the development of knowledge?

Possibility and Certainty

Consider the following example.

I am standing in front of a billiard table. Between me and it, however, there is a screen that prevents my seeing the middle third of the table's surface. As I watch, I see a white ball travel rapidly across the table and disappear behind the screen. It then emerges on the other side and continues on its way. But that does not always happen. Sometimes a red ball emerges. The first time this happens, I merely note the fact. With further experience, however, I begin to wonder if the ball emerging is more often red than I would expect if this happening were simply fortuitous. In consequence, I begin to suspect that there may be some connection between the entry of the white ball and the emergence of the red. And the more often I see this happen, the stronger my suspicion becomes.

By now, my curiosity is aroused and I cannot rest until I have given myself an explanation—a hypothesis—about the apparent sequence of events. One possibility that occurs to me (or might, if I were an untutored person who knew nothing about billiards) is that something happens behind the screen to change the color of the entering ball from white to red. After all, I have seen apples change from green to red on ripening and litmus solution change color when acid is added to it. On the other hand, I might think that, perhaps, there might be a stationary red ball behind the screen and that, if the entering white one makes contact with this, it could impel it to move into the open. I might then recollect that each time before the red ball appeared I had heard a sharp click, such as billiard balls make if they knock together. Furthermore, that it was only if the white ball was traveling in a particular direction that the red subsequently emerged. I now begin to incline toward the second hypothesis. As I cannot see what goes on behind the screen, however, I cannot say which, if either, of these explanations is correct.

But suppose that the next time I see a white ball crossing the table I remove the screen. There I see a stationary red ball (the existence of which was previously unknown to me) and, as I watch, the white ball strikes it. I then hear the crack of the impact, see the red ball shoot smartly away and the white come to rest. I now possess factual information regarding the whole sequence linking the entry of the white ball to the emergence of the red and, consequently, am relieved of any need to speculate about the connection between these two events.

So my explanation of how the result in question comes about is progressively translated from the status of a suspicion to that of a possibility, from that of a possibility to that of a probability. When I have actually observed the complete sequence, however, it passes into another category altogether, that of a certainty. It has ceased to be a hypothesis and become a matter of fact.

Possibility and Certainty

Simple as this example may be, it illustrates several important points regarding the way human understanding of the causation of natural events develops.

III

In the foregoing example, it is evident that the frequency with which, after the entry of the white ball, a red one emerges from behind the screen does nothing to establish the validity of any explanation of why it does so. It merely increases the likelihood of there being some connection between the two events and so strengthens my motivation to discover this. No matter how often we see one thing follow another, that is no proof that they are causally related. Only direct observation that two events are actually linked will suffice to do this. Once we see that this is so, however, we are in a position to say for a fact that there is a connection between them and to generalize accordingly.

We can then say, for instance, that when any billiard ball is traveling in such a direction that, if it continues on its way, it will collide with a free-standing stationary one, the former will slow down or stop and the latter start moving. For, if there are any human beings alive to witness what ensues, after sighting a billiard ball traveling along such a trajectory, the course of nature will not have changed and the sequence of events will conform to that revealed by previous experience.

Scientific Knowledge and Philosophic Thought

But we can go farther.

Billiard balls are not the only inanimate objects that can become endowed with the property of motion. Exposed to the force that produces movement, any object that is not rooted to the spot can become a vehicle for this. Taking account of other objects in like situations, we can accordingly generalize more widely. We can say that, if a moving object of any kind makes contact with a similar free-standing, stationary one, it will transfer some of its property of motion to the latter. We can thus infer, on seeing any moving object, what would be the consequences of its colliding with another, even if we have never met such an object before, let alone seen it in collision with anything. Equally, if we see a previously stationary object begin to move, we can infer that something has made it do so. In short, given such a generalization, we can recognize the implications of a particular object for other things and, hence, the contribution it would make to the outcome of any sequence of which it was a part.

All this, however, is subject to one overriding proviso. We can only formulate valid generalizations about the bearing of one particular thing on another if our information regarding the chain of events which links them together is complete. But can we ever be sure that it is? Or is the position that, as in the case of descriptive generalizations, explanatory generalizations also can never be more than approximately valid?

There are grounds for thinking that this is not so.

Possibility and Certainty

Every sequence has a beginning and an end. Before its existence can even be suspected, observation must already have revealed its first and last items. Furthermore, every item in a sequence, after the first, is a consequence of that which precedes it; every one, before the last, an antecedent of that which follows. Accordingly, when knowledge has developed forward to the point at which the last item identified implies that the next will be the second of the two items initially observed, or backward to one that equally indicates that the item preceding it was the first of this pair, we can be sure our information regarding this particular sequence is complete. We then know for a fact that there is, albeit through a succession of intermediaries, a connection between the two events we had originally observed. In consequence, we can be certain that, whenever we see the first, the other will follow unless something untoward has happened to break the continuity of the sequence of things that ordinarily links them together.

And, why can we be sure? Because in sequences we are seeing the manifestation of forces that govern the course of nature. Unless, therefore, these have changed (in which case nobody would be alive to witness what happens), the same forces, acting through the agency of the same objects, cannot do otherwise than produce the same effects.

It would seem, therefore, that explanatory gen-

eralizations are not subject to the same limitations
as descriptive. Of necessity, the latter are limited by
experience. But the laws of nature are universal in
their application. In consequence, generalizations
involving those laws that are valid in a limited
context will be universally so.

Admittedly, the example we have hitherto been
considering is artificial. As such, it may appear too
slight to support a conclusion of this magnitude.
Accordingly it might be as well to look at some
examples from real life.

V

From the earliest times, men have seen that lightning precedes thunder. Furthermore, they have noticed that the interval between seeing a flash of
forked lightning and hearing a peal of thunder
increases with the distance from the place where the
lightning comes to earth. Knowing, from other
experience, that they see things at a distance before
the sound they make is heard, men could not fail to
conclude that lightning is the "cause" of the thunder. It was not, however, until some two centuries
ago, that any progress was made toward establishing the connection between the two events. Then, as
a result of Benjamin Franklin's classical experiments, it was revealed that lightning was due to the
discharge of electricity that had accumulated in the
clouds. But it was not until much later that it was

shown that the temperature of a lightning flash was many thousands of degrees centigrade. In consequence, the column of air through which it passes expands at supersonic speeds. Inevitably this gives rise to waves in the surrounding atmosphere which, impinging on the human eardrum, are registered as sound. And sound waves traveling much more slowly than light, this necessarily entails that the flash of the electrical discharge is seen before the disturbance in the surrounding air is heard.

Now consider an example from the biomedical field.

Because of its devastatingly high mortality and its tendency to spread in human populations, cholera has long attracted men's attention. It was not, however, until the middle of the nineteenth century that any substantial progress was made toward explaining these observations. Nevertheless, this did not prevent people from fabricating explanations to account for outbreaks of the disease. At first, these were regarded as a manifestation of the wrath of supernatural beings angered at the disregard by men of their injunctions. Later, as due to the inhalation of some hypothetical poison emanating from swamps, stagnant water, dirty dwellings, or putrefying corpses during the heat of the summer. On this latter basis, many ingenious explanations were proposed. None, however, quite met the case.[1] Then, during the last cholera epidemic in London, a physician, John Snow, noticed that, in his district, patients with this disease were largely confined to

Scientific Knowledge and Philosophic Thought

persons who obtained their drinking water from a
particular pump. Moreoever, he saw that a cesspool
with broken walls adjoined the well on which this
pump drew and that the drains from a house containing a case of cholera led into this. So, he removed the handle from the pump and the epidemic
in his locality stopped.

Thus the impasse to the progress of knowledge
about this disease was broken. Before this time, the
only firm fact was the existence of an epidemic
disease with a particular combination of characteristics. Now a further fact had been established: that
there was an association between an outbreak of
cholera and a water supply contaminated by the
excreta of persons suffering from this disease. It was
not, however, until some thirty years later that any
progress was made toward clarifying the sequence
that linked these two events. Then Koch isolated the
causative organism of the disease and so took the
first step toward establishing the connection between them.

Looking at the current scene in scientific research, we can see similar sequences in every stage
of completion. Take the apparent association between cigarette smoking and lung cancer. It is in the
highest degree probable that there is a connection
between the two. But the nature of this has not yet
been factually established. We are in the same position as when the association between outbreaks of
cholera and a contaminated water supply had been
demonstrated but Koch had not yet discovered the

Possibility and Certainty

cholera vibrio. That, however, constitutes no reason for our not acting on the basis of the probability we do have—as we did, with marked success, in regard to cholera.

VI

Generalizations deriving from observation of the order in which things succeed each other in nature lie, therefore, at the basis of man's ideas about causal relationships. As such they are the foundation of all his understanding. For, to what is this directed but to elucidating how things come about or what they portend? It follows accordingly that the progress of human knowledge depends primarily on the ability of man to increase his stock of generalizations of this kind.

We thus come back to our essential proviso. An explanatory generalization can only be authentic insofar as the observations on which it is founded are, not only factually correct, but complete. In the case of a relationship in which one thing has a directly observable effect on another this presents no difficulty. In the case of one in which two events are related through a sequence of intermediaries, however, the position is different—for some, or all, of these may not as yet have been observed.

This is the situation that ordinarily confronts scientists at the outset of an investigation. The question is to what extent, if any, can they remedy deficiencies in their factual information? Must they wait passively for chance experience to reveal what is missing? Or can they take active steps to obtain this?

Looking at the preceding two examples, it will be seen that progress toward solving the problem of whether two things were causally related depends on asking if something else was present which, if it were, would go at least some way toward explaining what happened. But, how could anybody have any inkling of what might be present, if what happened in between the first and second events noted is outside their experience? The answer is, of course, that they do what people always do in such circumstances. They call upon their imagination to fill in the gaps in their factual information and thus enable them to construct a hypothetical sequence to account for the association in question and, hence, give them an indication of the kind of thing to look for. On the face of it, this is a hazardous procedure. We should do well to ask, therefore, to what extent, if any, do the fabrications of the imagination have any foundation in fact?

SIX

Imagination &
Credibility

That men can imagine things of which they have no experience is beyond doubt. But is the human mind completely free in this respect? Is the imagination entirely undisciplined, apart from the requirement that its fabrications shall not be self-contradictory? If so, is it not a delusion to believe that we can distinguish a concept that is merely a figment of the imagination from one that has a basis in reality? Even David Hume agreed that there is a difference between "the ravings of a madman" and "the sober efforts of genius and learning."[1] But how did he know? Given the premises from which "madmen" start, their contentions are often entirely logical. Yet, if we cannot make this distinction on logical grounds, on what grounds can we do so?

To the early philosophers of science, like Francis Bacon, this question presented no difficulty. Believ-

ing, as they did, in the feasibility of inductive thinking, they took it for granted that, provided the
observer had accumulated the requisite facts regarding his problem, his imagination would "automatically" be led to form a realistic hypothesis
about it. That it was impossible to produce a logical
justification for this belief did not worry them. They
were content to rest their case on their observation
that the hypotheses at which they thus arrived
tended to work. To many modern philosophers,
however, this failure to provide a logical basis for
inductive argument constitutes an insuperable objection to accepting the traditional view. In consequence, they have been driven, *faute de mieux,* to
regard hypotheses as the products of intuition, or
inspiration, rather than of rational thought.

And yet, even the proponents of the modern view
find themselves forced to concede that man is not
entirely free to accept whatever conjecture his imagination throws up; that before submitting a hypothesis to the test of refutation by further observation,
he has, in practice, some inkling of whether it is
likely or not. As Medawar put it: "If it is a formal
objection to classical inductivism that it sets no
upper limit to the amount of factual information we
should assemble, so also it is a defect of the hypo-
thetico-deductive (i.e. the 'conjecture-refutation')
scheme that it sets no upper limit to the number of
hypotheses we might propound to account for our
observations. . . . In real life, of course, just as the
crudest inductive observations will always be lim-

ited by some unspoken criterion of relevance, so also the hypotheses that enter our minds will be plausible and not, as in theory they could be, idiotic. But this implies the existence of some internal censorship which restricts hypotheses to those that are not absurd and the internal circuitry of this process is quite unknown."[2]

But is that so? That something exists which imposes restrictions on flights of imagination was tacitly conceded both by Hume and Popper. It is instructive, therefore, to see how they tried to deal with this problem.

II

"We can," says Hume, "in our conception join the head of a man with the body of a horse, but it is not in our power to believe that such an animal has ever really existed. It follows, therefore, that the difference between fiction and belief lies in some sentiment or feeling which is annexed to the latter, not to the former, and which depends not on the will, nor can be demanded at pleasure. . . . Whenever any object is presented to the memory or the senses, it immediately, by force of custom, carries the imagination to conceive that object which is usually conjoined to it; and this conception is attended by a feeling or sentiment different from the loose reveries of the fancy. In this consists the whole nature of belief. For as there is no matter of fact which we

Imagination and Credibility

believe so firmly that we cannot conceive the contrary, there would be no difference between the conception assented to, and that which is rejected, were it not for some sentiment which distinguishes the one from the other."[3]

So, despite his contention that "nothing is more free than the imagination of man," Hume conceded that it is only free within certain limits. These, he thought, were determined by the "force of custom," or "habit," as he also calls it.[4] It was this, in his view, that led the imagination, whenever any object is presented to the memory or the senses, immediately to conceive of the object with which it is usually conjoined. "It is that principle alone," he writes, "which renders our experience useful to us and makes us expect, for the future, a similar train of events with those which have appeared in the past."[5] Yet, how could we have got into the habit of expecting one particular object to be conjoined with another equally particular one, unless previous observation had revealed to us that it usually is? In that case, does it not follow that our inability to believe in the existence of an animal in which the head of a man is joined to the body of a horse is because such a beast is outside our experience?

Of course, Hume recognized that, in real life, past experience did impose restraint upon his imagination. Committed as he was to the belief that man could not count on the future resembling the past, however, he was theoretically unable to accept that compatibility with previous experience provides any test of the validity or otherwise of our imaginings.

A comparable difficulty faced Karl Popper.

Because of his belief that hypothesis always precedes observation he was driven to postulate the existence of an inborn sense that impels men to see regularities in natural phenomena. But, as we have seen, there is a far simpler explanation; namely, that such regularities actually exist and are revealed to them by observation. In that case, does it not follow that this tendency to order observational data in certain ways, and not in others, is not inborn but is acquired as the result of experience, their own and that of others?

There does seem, therefore, to be a general, albeit tacit, agreement that the imagination is not entirely undisciplined; that, over and above it, there is something that restricts its freedom of action. For Hume this is the force of custom or habit which, by channeling the imagination to envisage only objects that are usually conjoined with what is present to the memory or the senses, limits the number of credible concepts it can produce. For Popper it is an inborn propensity that inclines men to see regularities in natural phenomena and, hence, to expect any concept they form to reflect these. For Medawar, it is something akin to an internal censor that suppresses implausible conjectures.

Despite their differences, all these suggestions carry the same implication; namely that, somehow, man has come to possess criteria which enable him to judge the likelihood, or otherwise, of any concept his imagination throws up. If this be so, it would go a long way toward explaining, not only the process of hypothesis formation but also that of induction. The question is, therefore, whether such criteria actually exist and, if so, what is their nature and origin?

SEVEN

Inference, Induction, & Intuition

When individuals are confronted by a familiar object —one of a kind about which they have already formulated a generalization—then, as David Hume said, a glimpse of one of its attributes will immediately induce in their minds the idea of that with which it is usually conjoined. This, however, is only half the story, for the attribute observed will equally firmly exclude the idea of anything with which it is not usually so joined.

It is the same in regard to sequences. The sight of one item will immediately induce the idea of that which usually follows or precedes it and equally exclude that of anything that usually does not. Thus, a flash of lightning immediately induces the idea of thunder; the sight of smoke, the idea of fire. In other words, as a result of the generalizations they form on the basis of their experience of the way things are ordered in nature, men build up a knowledge of what things are compatible and what are not; what follows or precedes what and what does not. Consequently, on seeing one they are able to infer with what it is ordinarily linked.

But generalizations differ greatly in scope. In the case of descriptive generalizations, the simplest are those that relate only to objects of a specific kind, for example, European swans. Next come those that comprehend groups of objects, each of which differs in some way from the others but all of which have a sufficiency of attributes in common to justify their being grouped together; for example, ducks. Thirdly are those that apply to all objects in a major category of natural experience, such as animals.

Take the hypothetical concept of a centaur, cited by David Hume, and consider the several generalizations required to exclude it. "We can," he wrote, "in our conception join the head of a man with the body of a horse, but it is not in our power to believe that such an animal has ever really existed." But why is it not in our power to do this? It is not merely because such a creature is outside our personal experience. Nor is it only that it is contrary to a generalization founded on all human experience, namely, that a man's head is always joined to a man's body. Rather it is because the concept of a beast in which the head of one species is joined to the body of another runs counter to a wider generalization borne in upon us by our whole experience of how things are organized in the animal kingdom.

Scientific Knowledge and Philosophic Thought

It is the same in regard to sequences.

Take our previous example of billiard balls. Our first generalization was that, if a moving ball makes contact with a free-standing stationary one, it will cause the latter to move. Our next was that if any moving object collides with a free-standing stationary one of the same kind, it will do likewise. Our third was that, even if the objects are different, if one is moving and makes contact with one that is not, it will impart some of its property of motion to the latter. Similarly, we can, by an effort of will, conceive that, if we let go our hold of a stone, it will fly upward. But we cannot believe that it will. Why? It is not merely because neither we, nor anyone else, have ever seen such a sequence of events. It is because of the wider generalization we have formed that anything heavier than air will fall to the ground if unsupported. Or again, we can conceive of a plant in which fruits precede flowers, and flowers precede buds. But we cannot believe it exists. Why? Because the concept of such a sequence is contrary to a wider generalization man has reached on the basis of his whole experience of the sequence of events in the vegetable kingdom.

II

Thus, as a result of his experience of the world around him, man acquires a series of generalizations about the way matters are ordered in nature.

Inference, Induction, and Intuition

Each of these gives rise to certain expectations: that such-and-such will be conjoined with so-and-so and not with anything else; that this will be followed by that, and not by some other thing. Inevitably such generalizations limit the number of possibilities that can be entertained regarding the implications of any particular observation and, consequently, the number of hypotheses that can be accepted about the experience in question. For the human mind can no more accept a concept that combines ideas of incompatible things than one that is self-contradictory.

It could well be, therefore, that generalizations about the way things tend to be ordered in nature are the source of those "unspoken criteria of relevance" which Medawar saw research workers must possess if they are to select from a medley of observations those that are relevant to their problem. Furthermore, such generalizations would themselves act as "a form of internal censorship" such as he saw was required to explain why "the hypotheses that enter our minds will, as a rule, be plausible and not, as in theory they could be, idiotic."[1] Moreover, generalizations of this kind would also account for that tendency to find "regularities" in natural phenomena to which Popper drew attention and which, mistakenly in my opinion, he thought was inherent in man.[2]

It would seem, therefore, that generalizations about the way matters are ordered in nature lie at the basis of inductive thought. Given such, we can envisage the antecedents and consequences of individual items of information about a problem and arrive at concepts that enable us to appreciate its overall significance. In other words, man is a pattern-forming animal, and the patterns into which he orders the data supplied to him by his senses or memory are dictated by his experience of how things in nature fit together. Accordingly, it is from this point of view that we should look at the process of hypothesis formation.

III

Faced by a problem, man feels impelled to solve it. To this end, he first seek to order his observations of its attributes into a pattern that will enable him to recognize it when he sees it again and, in the light of the extended experience he thus acquires, to generalize about the particular combination of properties which characterizes things of this kind. But this only tells him that things of this particular kind exist. To understand how they came to do so, or what they portend, a further generalization is needed; one that will tell him what precedes or follows the thing in question.

On first encounter, observation rarely supplies an individual with the information required for this purpose. If it did there would be no problem, for, given information about a thing's immediate antecedent and sequel, its implications would be evident. At the outset, however, men's observations about any phenomenon are usually limited to those

Inference, Induction, and Intuition

that relate to its distinguishing features. Indeed, this may long continue to be the case. In such circumstances, it might seem, therefore, that we have no choice but to wait on chance experience to provide us with more information. But that is not necessarily so. Although, in the absence of a generalization about its implications, the sight or idea of a particular thing cannot induce in the mind any idea of what precedes or follows it, what it can do is to call to mind ideas of things it more or less resembles. And for some of these, at least, we may possess generalizations as to the kind of thing that does so in their case. In the light of the analogies with which he is thereby furnished, the observer can then envisage a number of possible antecedents or sequels to what he has observed. Certain of these he will discount at once on the grounds that they conflict with broader generalizations he possesses about the order in which things succeed each other in nature. Those that are not so disqualified, however, he will accept and, on this basis, construct a series of tentative explanatory generalizations—hypotheses—about the possible causation or consequences of the matter in question.

This is, in my opinion, the essence of the process of hypothesis formation.

Should observation now confirm that one of the possibilities thus envisaged is correct, the investigator will have obtained a further item of factual information about the sequence in question.

Theoretically, it is possible by following the above procedure to work back step by step from the item that first aroused the observer's curiosity to that which set in train the sequence that brought it about. Alternatively, if the item first observed was that initiating the sequence, the observer could similarly work forward to the outcome of this.

Scientific research workers, however, are not always restricted to working backward (or forward) in this way. It may be that the thing they are anxious to understand seems to follow on something else more often than would be expected if this were due to chance. Thus thunder follows lightning. There is an unduly high incidence of lung cancer among cigarette smokers. In neither case is there, at first, any evident connection between the things observed. Only a possibility that there might be one and that what is being seen are the first and last items of an as yet undiscovered sequence. Acting on this assumption, research workers can now approach their problem, so to speak, from both ends, working back from that which follows and forward from that which precedes. This considerably simplifies their task. Now their search is directed to a specific goal. Now their aim is to bridge the gap between two probably related observations. Thus oriented, the danger of either approach drifting off course is diminished. Moreover, the risk of research into a particular problem being halted by lack of suitable analogies or by technical difficulties is ob-

Inference, Induction, and Intuition

viously less than when there are two approaches to it rather than one.

But whatever approach is adopted there is one paramount requirement. Every hypothesis must be rigorously checked against further experience *before* another is projected. If any one fails to measure up to this test—in the sense that what it predicts should occur is not confirmed by further observation—then it must be ruthlessly discarded. This is because, being derived from analogy, hypotheses can rarely be more than approximately correct. In consequence, if a succession of hypotheses is built up on the basis of a succession of unconfirmed predictions, the likelihood of error is progressively compounded. It is this that accounts for the scientist's distrust of sustained flights of abstract thought, however logical they may be.

Looked at from a practical point of view, the above may seem a counsel of perfection. Obviously, if any hypothesis needs to be confirmed by further observation before the next is formulated, progress depends upon our having the means at our disposal to make the observation required. This is true and it accounts for the effort scientists devote to devising new techniques for increasing their factual information. Nevertheless, if progress is held up in this way, an investigator need not immediately abandon hope. He can assume that what his unconfirmed hypothesis predicts is correct. On this basis, he can then construct a further hypothesis and so on, until he arrives at one that is technically possible to

Scientific Knowledge and Philosophic Thought

confirm or refute. If observation does in fact confirm this, he is then in a position comparable to that of a man seeking to bridge the probability gap between two apparently related happenings, save that, in this case, his motivation to find means to do so is fortified by the knowledge that, at least in theory, a connection exists.

It seems, therefore, that there is nothing mysterious about the process of hypothesis formation. Why, then, has it been felt necessary to postulate a mysterious factor like intuition in order to account for it? In my opinion, the short answer is, primarily because of the peculiar way in which the products of inductive thought sometimes present themselves to the conscious mind of the individual.

IV

Man has long been aware of occasions when, defeated by a seemingly insoluble problem, its solution has suddenly, and apparently fortuitously, come into his mind. Understandably, the high degree of relevance of the concepts that thus erupt to the problem that has been engaging his attention (coupled with the fact that they do so without conscious effort on his part) has led him to regard such experiences as in a class by themselves. It is not surprising, therefore, that he has come to believe that concepts of this kind are the product of a process outside that of ordinary thought. Nor is it surprising that, aware of his own inadequacy, he should have felt gratified by the receipt of such unexpected help and disposed to credit it with special authority. As F. C. Bartlett said when discussing this phenomenon, "There comes with it a rather sharp thrill of pleasure, which I do not get from other ways (of thinking) and generally a quite peculiar feeling of certainty which, alas, is not always justified when it stands up to further tests."[3]

Now, however, when we know how much of an individual's thought goes on below the level of consciousness, is there any need to postulate a special process to account for such experiences? Cannot the genesis of so-called intuitive concepts be explained in the same way as that of any other man's imagination throws up in the course of his endeavors to order his data about a problem into a pattern consistent with the way experience has taught him things fit together in nature? Only in one respect do such concepts differ from those he arrives at by conscious thought: They are the products of mental activity at levels which do not ordinarily obtrude on his attention.

Anybody who has engaged in teaching has seen "intuition" in the making. At one time, I taught medical students during their clinical years. On their first entering the hospital wards, one had to take them laboriously through the findings on a patient and lead them to see that these were consistent with a descriptive generalization regarding a particular illness and not others. Within a few

months, however, they were beginning to do this for
themselves. At first, they proceeded slowly, spelling
out every step; later, with increasingly "unthink-
ing" speed and accuracy. By the time they came to
be medically qualified they were reaching a stage
when, in regard to most ordinary illness, they
"automatically" excluded certain diagnoses and
seriously considered only certain others. Thereafter,
with increasing experience, this unspoken, compu-
terlike ability develops progressively until, in the
case of experienced physicians (as every teacher of
medicine knows), it is often only by making a delib-
erate effort that they can recall the steps by which
they arrived at a particular diagnosis.

The same process is seen at work in the research
field. Perhaps the best known example is that involving Charles Darwin and Alfred Russell Wallace.

For twenty years Darwin had been quietly ac-
cumulating evidence in support of his long-
cogitated theory on "the origin of species by means
of natural selection." Across the other side of the
world, Wallace was recovering from fever on a
remote island in the East Indies, following a pro-
tracted tour collecting the flora and fauna of those
regions. Suddenly, out of the blue, the thought
flashed into his mind that the distribution of different kinds of plants and animals could be explained if
these had a tendency to depart indefinitely from the
original type, but only those departures survived in
any one place which were suited to the conditions
prevailing there. He immediately put his thoughts

Inference, Induction, and Intuition

on paper and sent them to Darwin. Darwin was dumfounded, for, he said, "this essay contained exactly the same theory as mine."[4] Yet he knew Wallace could not have heard of this, for it was unpublished. The upshot was that both Wallace's essay and an abstract by Darwin of his proposed book were published in the *Proceedings* of the Linnean Society for 1858.

There are many other examples of this kind of happening, not all of which had such a happy outcome. Only too often such "coincidences" have led to acrimonious disputes regarding priority of discovery. But is it surprising that, given a similar knowledge of the facts and similar background generalizations, trained scientists in different places should, more or less simultaneously, arrive at similar explanatory generalizations about them? Or that, just as an experienced physician may reach a diagnosis with little or no conscious thought, so an experienced research worker may develop a "hunch" about a scientific problem?

V

It would seem, therefore, that the traditional view of the way in which scientific knowledge is developed—although it leaves much unsaid—is essentially on the right lines. In particular, this is so in regard to the problem of hypothesis formation. The question is, Do similar considerations apply to the development of knowledge about matters that are commonly considered outside the province of science? To answer this question it is necessary to look at an example of thought from nearer the other frontier of that No Man's Land between science and theology.

EIGHT

Properties & Values

Perhaps the most influential book on moral philosophy published during this century was G. E. Moore's *Principia Ethica*. Recognizing that the purpose of ethical studies is to inquire into what is good, Moore starts from the contention that "how 'good' is to be defined is the most fundamental question in ethics." For, he writes, "unless this first question is fully understood, and its true answer clearly recognized, the rest of ethics is as good as useless from the point of view of systematic knowledge."[1]

It is consequently somewhat disconcerting to find Moore saying, on the next page, "If I am asked 'how good is to be defined?' my answer is that it cannot be defined and that is all I have to say about it."[2]

This, however, is not as startling a statement as it might appear at first sight. All Moore is saying is that 'good' is one of those simple ideas that cannot be analyzed into simpler ones and, hence, cannot be defined in terms of anything else.

To illustrate his point, Moore takes the example of 'yellow'. Obviously, one cannot convey the idea of 'yellow' to a person who is born without sight. It is a unique sensory impression. It follows, therefore, that, if an individual is born without the necessary sensory apparatus to register this property in external objects, it is impossible to describe it to him or her in terms of other sensations.

But is not this a mistaken analogy?

Provided their sight is not impaired, there is no disagreement between people as to what is yellow and what is not. Consequently, confronted by something that produces this particular sensation, they all immediately label it with the term in their language which denotes this.

But tastes differ. Some women, for instance, like yellow dresses. Others do not. Neither, of course, disputes that the garment is yellow. That is a matter of fact. But their reactions to it are different. To one, a yellow dress is beautiful; to another, ugly. There would seem, therefore, to be more than an element of truth in the old adage, "Beauty is in the eye of the beholder."

In principle, this consideration applies to all value judgments. What men consider 'good', and what they consider 'bad', differs from man to man, from group to group, and in the same group at different periods in its history. To one individual sweet dishes are repulsive; to another, delectable.

To some a particular book is good; to others, bad. Conduct that is highly commendable to the members of a terrorist organization may outrage law-abiding citizens. Beliefs that are accepted without question at one time may be wholly rejected at another. In short, when a person says that something is 'good', he or she is making a statement, not about it, but about his or her reaction to it. Any attempt to define 'good' on the basis of a property of the things to which people apply this term would accordingly be as mistaken as if one were to try and define 'beauty' in terms of a color.

This is the substance behind what Moore called "the naturalistic fallacy." And, as long as he restricted the use of this term to the above error, it is acceptable, save insofar as it tends to imply that this particular fault is one to which "naturalistic philosophers" are unduly prone. Unfortunately, however, Moore failed to distinguish between terms such as 'yellow', which relate to a property of a thing itself, and those such as 'beautiful' which refer to the observer's reaction to it. In consequence, he was led to dismiss arguments based on both as naturalistic fallacies. It is hardly surprising, therefore, that he was unable to define 'good'; by this oversight, he deprived himself of the means of doing so.

We should accordingly recognize that 'good' is not a property of things in themselves but an attribute with which individuals invest them in virtue of their own predilections. This is the essential distinction between properties and values. It is from this standpoint, therefore, that value judgments fall to be considered.

II

Looking at the terms people use when making value judgments, it will be seen that all denote either approval or disapproval. This holds even though, when talking of different things, they tend to use different words. In passing judgment on a work of art, for instance, they will ordinarily say that it is either 'beautiful' or 'ugly'; on a solution to a problem, 'right' or 'wrong'; on a course of action, 'prudent' or 'imprudent'; on human conduct, 'moral' or 'immoral'. The words 'good' and 'bad', respectively, comprehend all such terms. It would seem, therefore, that, on first analysis, 'good' is simply that of which the speaker approves; 'bad', that of which he or she disapproves.

But human beings are not only capable of distinguishing what is (in their estimation) good from what is bad. They can also distinguish the extent to which anything is one or the other. This they do on the basis of the strength of the approval or disapproval they feel toward it. If, for example, they approve of one thing more than another, they will describe it as 'better'; if more than any other, as 'best'. Similarly, in grading things they regard as 'bad', they will use the terms 'worse' or 'worst'. This applies irrespective of what is under consideration—be this an object, the solution to a problem, a course of action, or a matter of conduct—and whether it is present to the observer or not. For the idea of a thing is as conducive as its presence to the arousal of these sentiments.[3] Accordingly, by taking account of the relative strength of the approval or disapproval they feel toward two different things, or their ideas of these, men can evaluate one against the other, even though, in themselves, neither has any actual property in common.

The value judgments men are thus led to make play a major part in shaping their motivation and, hence, in determining their conduct.

But why do men approve of some things and not of others? In other words, what deeper significance attaches to their value judgments?

III

There are certain basic needs that human beings, like other living creatures, must satisfy if they, and through them their species, are to survive. They must obtain enough to eat. They must protect themselves from adverse factors in their circumstances, including those arising from the activities of others of their kind. They must ensure their ability to reproduce. To this end, all are endowed with inherent traits that impel them to take action to satisfy their needs in these respects. Such traits reveal themselves to the conscious thought of individuals by the emotions with which they inspire them: pleasant if the need in question is being satisfied; painful if it is not.

But man is not only an individualist. He is also a social being. As such, he has additional needs.

It is now more than a century since Charles Darwin published his book *The Descent of Man*. In this book he clearly distinguished between those species of animal that operate alone or, at most, in family units, and those that live in communities and operate collectively. The latter, he saw, must have evolved an additional trait which inclined them to act in concert and refrain from mutual aggression. Further, he saw that man, being, at least in part, a social animal, such a "social instinct" would account for what was ordinarily called his "moral sense."[4]

Despite Darwin's by then immense prestige, however, his views made singularly little impression. Not only Herbert Spencer but even Thomas Huxley, when writing on ethics, failed to appreciate their relevance. But they were not wholly forgotten. It was not, however, until some forty years later that any substantial progress was made toward developing them further. Then, in the first decade of this century, Wilfred Trotter published his essays "Herd Instinct and Its Bearing on the Psychology of Civilized Man."[5] In this the consequences of man's being a gregarious animal were translated into terms of individual psychology. Unfortunately, by this time, the tide of popular opinion was beginning to run strongly against explanations of human conduct which attributed any significant influence on man's motivation to his inherent attributes, as distinct from his environmental circumstances. In consequence, after an initial period during which, fortuitously, because of the First World War, it attracted attention, Trotter's thesis has been largely forgotten. Nevertheless, it remains the most noteworthy attempt yet made to arrive at an objective explanation of value judgments and, as such, is still deserving of attention.

IV

In ordinary life, people customarily distinguish between those of their activities that are only formulated after conscious deliberation and those they undertake without any forethought. The former they loosely call "rational"; the latter, "instinctive." In the last decade of the nineteenth century, William James, the psychologist, made some pertinent observations on the way "instincts" that bear on the satisfaction of an individual's personal needs, manifest themselves in his conscious thought.

It is, he pointed out, a peculiar feature of any idea that relates to the satisfaction of a biological need to appear to the individual as self-evident. Of course, a man likes things that satisfy his hunger. Of course, he dislikes anything that threatens his security. Of course, he is attracted by a pretty girl. And the same applies to his ideas about the conduct required of him if he is to gratify his wishes in any of these respects. Of course, he will take measures to obtain food. Of course, he will take steps to defend himself against possible attack. Of course, he will seek to develop his acquaintance with an attractive woman. For the individual concerned, the connection between the wish to be realized and the conduct required to do so is, in principle, as James said, "absolute and *selbstverstandlich,* an *a priori* synthesis of the most perfect sort requiring no proof but its own evidence."[6]

It was Trotter's particular contribution to extend these considerations to cover man's social propensities. Starting from the point that no human group could operate collectively unless its members thought similarly on matters of common concern, he drew the obvious conclusion; namely, that, in the course of his evolution, man must have developed an inherent trait that predisposed him to do so. But beliefs differ from group to group. Nevertheless, despite these differences, those prevailing in any one are regarded by its members as "axiomatically obvious propositions" which it would be foolish or wicked to question. To suggest that the members of different groups possess different inherent traits, each tailored to a particular idea, and that this accounts for the great diversity in all past, present, and future human beliefs, would obviously be absurd. On the other hand, such differences could readily be accounted for if man was endowed with an inherent trait that impelled him to subscribe to any belief that prevailed in the group to which he belonged and to adopt this as his standard for evaluating anything to which it relates.

In Trotter's time, social studies were still in their infancy. Even so, they were already producing data that were consistent with his views. Durkheim, for instance, from his investigation of primitive religions, was being forced to recognize the influence of group belief on individual conduct.[7] Again, in his classical studies on suicide—perhaps the most individual of all human actions—he was similarly driven to see that "the moral constitution of society establishes the contingent of voluntary deaths."[8] The upshot was that he felt compelled to conclude that "society is the source of all morality."[9] Equally, Piaget, investigating moral development in children, came to see the basis for this in the individual's "unilateral respect" for the customs and beliefs of those with whom he associated.[10]

And even in the sphere of moral philosophy there were not lacking traces of similar considerations.

David Hume, as would be expected, clearly grasped the essential difference between purely intellectual concepts and beliefs. "It is evident," he said, "that belief consists, not in the nature or order of our ideas, but in the manner of their conception and their feeling to the mind."[11] Later, Moore's predecessor, Henry Sidgwick, came close to seeing the significance of this difference. Writing some thirty years earlier, he drew a sharp distinction

Properties and Values

between what he called "prudential" and "moral judgments." There was, he saw, a qualitative difference in his feelings in the two cases. "The peculiar emotion of moral approbation," he write, "is, in my experience, bound up with the conviction, implicit or explicit, that the conduct approved is really right—*i.e.* that it cannot without error be disproved by any other mind. If I gave up this conviction because others do not hold it, or for any other reason, I may no doubt retain a sentiment prompting me to the conduct in question or—what is perhaps more common—a sentiment of repugnance to the opposite conduct; but this sentiment will no longer have the special quality of 'moral sentiment' so called."[12] And he concludes, "So far from being prepared to admit that the proposition 'X ought to be done' merely expresses the existence of a certain sentiment in myself or others, I find it strictly impossible so to regard my own moral judgment without eliminating from the concomitant sentiment the peculiar quality signified by the term 'moral.' "[13]

These were pertinent observations. And they became even more so when he went on to inquire into their significance. "Approbation, or disapprobation," he says, "is not the mere liking or aversion of an individual for certain kinds of conduct, but this is complicated by a sympathetic representation of similar likings or aversion felt by other human beings." "This, however," he goes on, "is partly because our moral beliefs commonly agree with those of other members of our society, and, on this

agreement depends to an important extent our confidence in the truth of these beliefs."[14]

Unfortunately, however, Sidgwick did not pur-
sue this line of thought. Had he done so, it could
hardly have failed to bring him into contact with
that of his contemporary Darwin on the biological
significance of moral emotions. In that case, subsequent thought in moral philosophy might have
taken a very different course. In particular, it might
then have been appreciated that when Moore—or
any other person—said, "If I am asked what is
'good', my answer is that good is good and that is
the end of the matter,"[15] he was making a psychological statement, and that, accordingly, the basic
problem in ethics is that of the psychobiology of
human motivation.[16]

V

It would seem, therefore, that any idea that comes to
be held in common by a group of men will, ipso
facto, acquire an ascendancy over their thought
processes. Thereby it is translated from the status of
a concept to be considered in the light of the evidence to that of a belief which it is morally incumbent
upon them to accept. This is revealed by the conviction with which it now inspires them, the guilt
they feel if they fail to heed its promptings, the
opprobrium they visit on any of their fellows who
deviate from its teachings, and the comfort they

Properties and Values

derive from knowing that they, themselves, are conforming to its requirements. For it is shared belief that lies at the basis of all social cohesion, and it is the function of man's inherent traits as a partly social animal to ensure that the beliefs prevailing in any group to which he belongs are endowed with the requisite emotional authority to ensure their acceptance and implementation by its members.

We should accordingly recognize that any idea that ministers to the promotion of social cohesion within any human group will appear good to its members; any that does the opposite, bad. But it is a fact of life that beliefs differ from group to group. To see this one has only to recall the vehemence with which the members of one political party will condemn the policies of those in another, or the bitterness with which the members of one religious sect will denounce the beliefs of those of a different persuasion. To impugn the sincerity, or intelligence, of the proponents of these different views would be naive. In consequence, any system of thought in ethics which starts from a proposition that neglects to take due account of these facts, and their implications, contains within itself the seeds of its own failure.

NINE

Science &
Philosophy

We may now attempt to answer the question from which we started, namely, Are there two kinds of problem—the scientific and the philosophic—each of which requires a different method for its solution; or are there two different methods for solving a problem, and according to which we use, we shall get a different answer?

Looking at the propositions from which the philosophical systems of Hume, Popper, and Moore start, it will be seen that each is based on a supposition. In Hume's case, his proposition that past experience provides no sure basis on which to infer the course of future events rests upon the supposition that the course of nature might change. Popper's proposition that hypotheses are the products of a process outside that of rational thought is based on the supposition that theory always precedes observation. Moore's proposition that 'good' is indefinable stems from his supposition that 'good', like color, is a property of what is observed.

These are all suppositions about matters of fact. Either the course of nature will change, or it will not. Either theory precedes observation, or it does not. Either 'good' is a property of what is observed, or it is not.

We may, therefore, define a proposition as an inference drawn from a supposition about a matter of fact. Accordingly, it is from this point of view that we should examine any proposition with which we are presented.

II

Take first Popper's contention that "the belief that we can start with pure observation alone, without anything in the nature of a theory, is absurd."[1] Certainly it is true that, if an individual has a theory about a problem, he or she will be predisposed to look in a particular direction for further information. Moreover, this is particularly so in regard to the most theoretically developed field of science, namely, physics. Had Popper paid attention to experience in other fields, however, he would not have been able to dismiss the part that accidental observation has played in the development of scientific knowledge in this summary fashion. He could then have seen that, even when observation is orientated by hypothesis, it is the unexpected and consequently unseekable element in a new finding, not the expected and seekable, that contributes to the advancement of understanding. As it was, he was led to conclude that observation is always preceded by theory and, as a result, driven to invoke the aid of intuition in order to account for the process of hypothesis formation.

The error in Hume's case goes deeper. He started from the proposition that *because* we cannot logically exclude the possibility that the course of nature might change, *then* man's belief that, on the basis of past experience he can infer from present happenings the course of future events, is unwarranted. This is, of course, a perfectly logical argument. As such, Hume apparently saw no cause to look into the matter further. But there was another possibility—namely, that the course of nature might *not* change—and this led, equally logically, to exactly the opposite conclusion. Nowhere, however, does Hume mention, let alone consider, this possibility. Had he done so, he would have been bound to give reasons for preferring the first of these alternatives. Inevitably that would have required him to consider the evidence in its favor. And, even in his day, relevant evidence was available, at least in regard to one of the forces that determine the course of nature, gravity. Newton had already published his *Principia*. Torricelli had suggested, and Pascal (with the aid of his brother-in-law) had shown, that the atmosphere has weight. It was common knowledge that men die unless they have air to breathe. Had Hume considered these facts, he could not have failed to see that, as long as man had existed, gravity must have been acting continuously. Moreover, that as long as there were any human beings alive to understand anything, it would not cease to do so. In other words, had Hume adequately investigated his suspicion that the course of nature might change, he would have seen that men could proceed in the sure knowledge that it would not do so in their time and consequently that, as far as the development of human understanding was concerned, man's belief that the future would resemble the past was well founded.

A similar, but a more serious, criticism applies to Moore's case, for his proposition that 'good' is indefinable rests on the mistaken assumption that 'good', like color, is a property of things. This arose from his failure to pay due attention to the fact that, as he knew, what is believed to be good differs from person to person, from group to group, and from time to time, whereas what is called 'yellow' does not. Had Moore done so, he would have seen that 'good' is not a property of what is observed but one with which the observer, because of his or her own predilections, endows it. To have conceded this, however, would have required him to take account of the biological and psychological factors that determine human opinion. And that Moore was not prepared to do. For, as he wrote, "What we want from an ethical philosopher is a scientific [sic] and systematic Ethics, not merely an Ethics professedly 'based on science,' "[2] which latter, so he says, "is inconsistent with the possibility of any Ethics whatsoever."[3] It is no wonder, therefore, that he inveighed so vehemently against those he called "naturalistic philosophers," for, in his eyes, they were betraying their art.

It thus appears that, despite their differences, the basic fault in each of these three systems of philosophy is the same. All start with propositions deriving from suppositions the factual basis of which has not been adequately explored. Perhaps, therefore, Francis Bacon was not as naive as is often supposed when he suggested that the first step in advancing knowledge about anything is to construct a "Kalendar" of all known facts that have a bearing on it. For, as he said, it is "in the nature of the mind of man (to the extreme prejudice of knowledge) to delight in the spacious liberty of generality . . . and not in the inclosures of particularity."[4] And if for "the mind of man" we read "the human intellect," this would seem a necessary precaution.

III

Despite opinion to the contrary, the intellect is not a discriminative instrument. It will process any data with which it is supplied, provided only it can form them into a consistent pattern.

Science and Philosophy

Clearly, however, no instrument could function in this way unless it had previously been programed to do so. It follows, therefore, that, before the intellect can operate, it must already possess information about what ideas or concepts are compatible and what are not. This information derives from the generalizations man has reached as to what kinds of things go together in nature and what do not. These are the generalizations that lie at the basis of the process of induction. For of what does induction consist but of fitting together of ideas and concepts into patterns that accord with the way experience has taught man matters are ordered in nature? Inasmuch as inductive thought involves joining ideas and concepts that are compatible and rejecting those that are not, it is, therefore, a logical process.

But this should not mislead us into thinking that logic provides us with a sufficient criterion for assessing the validity or otherwise of an intellectual production. Certainly any such product must be logical, in the sense that the ideas or simpler concepts out of which it is constructed must be compatible with one another. But that is only a requirement that qualifies it for further consideration. The test of its validity is that it shall not be contradicted by factual experience. Consequently, it is only by checking the products of the intellect against such experience at every step in its operations, from the construction of simple concepts out of simple ideas to the complex constructions it ultimately produces, that we can have any assurance that these correspond to the reality of that to which they purport to relate.

We should accordingly recognize that the reliability of the intellect as an instrument for advancing knowledge depends essentially upon the validity of the generalizations that program its operations. If these are well founded, so will be the concepts it produces. If, on the other hand, they are not, its productions will be correspondingly erroneous.

It is in this connection that propositions exert a decisive influence on the development of knowledge. For what is a proposition but a generalization based upon an assumption about a matter of fact? If the underlying assumption is correct, this is all to the good. If it is not, however, the intellect will but compound its errors. For, in order to ensure that the concepts it produces are self-consistent, the intellect is compelled to construe any further information with which it is supplied in such a way as to accord with the proposition that informs its operations in any particular instance.

Thus, Popper's proposition that hypotheses are the products of intuition rests upon the assumption that theory must precede observation. This assumption, however, would immediately become untenable if it were ever conceded that observation can be accidental. Accordingly, confronted by an apparently fortuitous observation, proponents of Popper's view are compelled to maintain, "The lucky accident fulfils an *a priori* expectation, however vaguely formulated it may have been."[5]

Hume found himself in a similar difficulty when he came to consider the problem of credibility. Why do men believe some concepts and reject others? The simple answer is that the former are consistent with the way experience has taught them matters are ordered in nature, whereas the latter are not. To have conceded this, however, would have required Hume to relinquish the proposition on which his whole system was founded. For human beings cannot believe in concepts that are open to doubt, and if they had no justification for believing that their future experiences would resemble their past, what basis had they for believing in one concept rather than another? Consequently Hume was driven to seek for another explanation. To this end he postulated that "as nature has taught us the use of our limbs without giving us the knowledge of the muscles and nerves by which they are actuated, so has she implanted in us an instinct [*sic*] which carries the thought in a correspondent course to that which she has established among external objects."[6] But if that is so, would man ever have to learn anything?

Moore's way of dealing with such difficulties was different. He did not attempt to explain them away. He simply sitgmatized any argument founded upon them as a "naturalistic fallacy" and dismissed it from consideration. At several places in his *Principia Ethica,* for instance, he discloses that he is aware that what people regard as 'good' at a particular time may not be so regarded at another. As his professed concern was to define 'good', one would have expected him to pay particular attention to this fact. It is accordingly important to see why he did not.

"By nature," he writes, "I do mean and have meant that which is the subject matter of the natural sciences or psychology. It may be said to include all that has existed, does exist, and will exist in time. If we consider whether any object is of such a nature that it may be said to exist now, to have existed, or to be about to exist, then we know that it is a natural object." He then goes on, "Of our minds we should say that they did exist yesterday, that they do exist today, and probably will exist in a minute or two. We shall say that we had thoughts yesterday . . . and, in so far as those thoughts did exist, they too are natural objects."[7] But in Moore's view, as we have already noted, ethics is not concerned with natural objects. "What we want from an ethical philosopher is a scientific [sic] and systematic Ethics, not merely an Ethics professedly 'based on science.' "[8] It follows, therefore, that it is a fallacy to take account of knowledge in the natural sciences in the context of ethics.

Whatever we may think of this contention, its practical import is evident. Armed with it, the holder is licensed to disregard factual information and confine himself to abstract speculation.

The conclusions to which the foregoing considerations point are clear. It is only to the extent that the intellect is disciplined by factual experience, at every step in its operations, that men can have any confidence in the validity of the concepts it produces. In short, concepts reached by abstract thought, however logical, have no title to validity save insofar as they are endorsed by factual observations.

Accordingly, it is only by rigorously investigating his factual experiences, and following his findings wherever they may lead, that man can provide himself with the requisite information on which to develop his understanding. Therein lies the difference between the scientific and philosophic approaches to the advancement of knowledge.

And yet, scientific knowledge could not have been developed so rapidly over the last three centuries had it not been preceded by philosophy.

IV

Fear of the unknown is deeply ingrained in human beings. In consequence, faced by anything they do not understand, they cannot rest until they have given themselves an explanation of it. This applies regardless of whether they are motivated to do so by an evident need or that sublimated sense of need we call curiosity. Provided they have the requisite information to solve their problem, they may succeed in doing so. If, however, they have not, they will grasp at any explanation rather than remain in suspense.

It is in this connection that philosophy has made an indispensable contribution to the advancement of human knowledge.

All explanations are generalizations about the way something comes about. As such they are constructed out of men's ideas about the implications of the various factors that are involved in this process. By introducing logic into speculation, and clarifying the difficult relationship between meaning and language, philosophy has done much to discipline man's thinking and, thus, to prevent his falling victim to intellectual misconstructions.

These considerations apply to the working up of ideas or concepts of any description. They are as relevant to those that derive directly from factual observation as to those that rest on supposition. As such, the procedures worked out by philosophers for handling ideas and concepts are as indispensable in the sphere of science as in that of philosophy.

Nevertheless, when it comes to assessing the validity of the products of intellectual activity, different considerations apply. For the products of the intellect can be no more valid than the ideas or concepts out of which they are formed. In other words, in this connection, supposition is no substitute for concrete information. It is accordingly only to the extent that man pays a docile attention to his factual experiences, and makes it his first concern to investigate these, that human knowledge can ever be more than opinion.

At any particular period of time, however, the factual information at our disposal is limited by the means available for obtaining this. And these may be wholly insufficient to supply us with all we require to solve many of the problems that beset us. Thus placed, we have no option but to do the best we can with the information we have and to fill in any gaps in this by speculation. Steadily, however, over the course of time, men have developed increasingly effective ways of investigating their experiences and so increasing the factual knowledge at their disposal. Of these the most notable are the experimental method and its extension, by prospective surveys, to the investigation of naturally occurring events. By the seventeenth century this more rigorous approach to the verification of factual assumptions and propositions was beginning to take shape, and modern science was born. At first, this approach was largely confined to problems in the field of men's material interests. Soon, however, it began to spread to others and to demonstrate its relevance to the advancement of understanding in fields that had been thought to be beyond its scope.

Looking to the future, the implications of these developments are clear. As scientific knowledge expands it will progressively take over more and more of that No Man's Land between it and theology. Indeed, this process is already in train. In the last three centuries, the physical sciences have wholly taken over that part of this territory which was once the preserve of Natural Philosophy. In the last hundred years, biology has revolutionized man's thinking about himself and his place in the universe. Today, consequent upon its increasingly scientific approach to social problems, sociology is beginning to impinge on Social and Political Philosophy. Now the invasion of the traditional province of Moral Philosophy by psychobiology cannot be long delayed. And, as Isaiah Berlin says, "systematic parricide is, in effect, the history of the natural sciences in their relation to philosophy."[9]

V

It thus appears that the difference between science and philosophy lies, not in the problems to which they are directed, but in the methods they use for solving these problems. Given the propositions from which science and philosophy start, both are equally logical. Where they differ is in their approach to the suppositions that underlie their propositions. To the scientist, a proposition is something to be investigated; to the philosopher, something (provided it is not illogical) to be accepted as a basis for thought.

If this be so, then the implications for the further development of human knowledge are profound.

Since man adopted the scientific approach to problems in his natural environment, he has made substantial progress toward mastering them. The same cannot be said, however, about his achievements in regard to those he generates for himself. Indeed, in respect of many of them, his position is little different from what it has always been. These are the problems that traditionally fall within the sphere of philosophy. It would seem, therefore, that, if man is ever to attain a degree of control over problems that derive from his own activities, comparable to that he now has over those imposed upon him by factors in his natural environment, it will only be by approaching them in a comparably objective way.

It would, however, be foolish to dismiss out of hand a proposition upon which men have long relied for guidance. Beliefs that have survived for centuries may be mistaken. But the experiences which lie behind them are seldom devoid of foundation. This applies to all propositions, including those that relate to matters which, at any particular period of time, are considered to fall outside the ambit of science.

Accordingly, what is needed now is not indiscriminate iconoclasm, but a systematic re-examination of the propositions from which thought in philosophy takes its start. To this end, the first requirement in respect of any particular proposition is to identify the supposition that lies behind it. Given this, it is then possible to investigate its factual justification and, in the light of the further information thus obtained, to appraise the validity of the proposition in question.

Scientific Knowledge and Philosophic Thought

It was no accident, therefore, that led the founding members of the Royal Society of London to take for their motto the words *Nullius in Verba*. And the results are there for all to see. Whether man will be able to bring himself to adopt a similar approach to those perennial problems which, with his increasing mastery of conditions in his natural environment, have now become the most important facing his species, is, however, an open question.

Notes

ONE   METHODS OF THOUGHT

1. Bertrand Russell, *A History of Western Philosophy* (London: George Allen & Unwin, 1946), p. 10.
2. David Hume, *An Inquiry concerning Human Understanding* (1748) (New York: Bobbs-Merrill, 1955), p. 61.

TWO   EXPERIENCE AND UNDERSTANDING

1. Bertrand Russell, *A History of Western Philosophy* (London: George Allen & Unwin, 1946), p. 685.
2. David Hume, *An Inquiry concerning Human Understanding* (1748) (New York: Bobbs-Merrill, 1955), sec. 4, pt. 2, p. 51.

THREE   OBSERVATIONS AND HYPOTHESES

1. Francis Bacon, *The Advancement of Learning* (1605), bk. 2, sec. VIII, para. 1.
2. Ibid., bk. 2, sec. XIII, para. 3.
3. Ibid., bk. 2, sec. VIII, para. 3.
4. Bertrand Russell, *A History of Western Philosophy* (London: George Allen & Unwin, 1946), p. 565.
5. Bacon, *The Advancement of Learning*, bk. 2, sec. XIII, para. 9.
6. Ibid., bk. 2, sec. XIII, para. 10.
7. Karl Popper, *The Logic of Scientific Discovery*, rev. ed. (London: Hutchinson, 1968), chap. 1, sec. 1, p. 27.
8. Ibid., chap. 1, sec. 2, p. 32.

9. Karl Popper, *Conjectures and Refutations,* 2d ed. (London: Routledge & Kegan Paul, 1965), p. 46.

10. Ibid.

11. Ibid., p. 47 (the italics are the author's).

12. H. H. Dale, "Accident and Opportunism in Medical Research," *British Medical Journal,* no. 2, 1958, p. 451.

13. P. B. Medawar, *Induction and Intuition in Scientific Thought* (London: Methuen, 1969), pp. 30 and 41.

FOUR   THE PARTICULAR AND THE GENERAL

1. David Hume, *A Treatise on Human Nature* (1738) (London: J. M. Dent & Sons, 1911), 1:107.

2. Ibid., p. 91: "We can at least conceive a change in the course of nature: which sufficiently proves [*sic*] that such a change is not absolutely impossible."

FIVE   POSSIBILITY AND CERTAINTY

1. An instructive account of the development of thought on the genesis of cholera is given in S. N. De's book *Cholera* (Edinburgh: Oliver & Boyd, 1961), chap. 1.

SIX   IMAGINATION AND CREDIBILITY

1. David Hume, *An Inquiry concerning Human Understanding* (1748) (New York: Bobbs-Merrill, 1955), sec. 1, footnote, para. 3. p. 33.

2. P. B. Medawar, *Induction and Intuition in Scientific Thought* (London: Methuen & Co., 1969), pp. 52 and 53.

3. Hume, *Inquiry,* pp. 61 and 62.

SEVEN   INFERENCE, INDUCTION, AND INTUITION

1. P. B. Medawar, *Induction and Intuition in Scientific Thought* (London: Methuen & Co., 1969), p. 53.

2. Karl Popper, *Conjectures and Refutations*, 2d ed. (London: Routledge & Kegan Paul, 1965), p. 47.

3. F. C. Bartlett, *Religion as Experience, Belief, and Action* (Oxford: Oxford University Press, 1950), p. 10.

4. Charles Darwin, *Autobiography* (London: Collins, 1958), p. 121.

EIGHT   PROPERTIES AND VALUES

1. G. E. Moore, *Principia Ethica* (1903) (Cambridge: Cambridge University Press, 1976), chap. 1, para. 5, p. 5.

2. Ibid., chap. 1, para. 6, p. 6.

3. Spinoza: "A man is affected by the same emotions of pleasure or pain from the image of a thing past or future, as from the image of a thing present." *Ethics* (1678) (London: J. M. Dent & Sons, 1910) pt. 3, prop. XVIII.

4. Charles Darwin, *The Descent of Man* (1871), 2d ed. (London: John Murray, 1887), chap. 4, p. 97.

5. These essays were published in the *Sociological Review* in 1908 and 1909. Later, Trotter expanded them in his book *Instincts of the Herd in Peace and War* (London: Ernest Benn, 1916).

6. William James, *Principles of Psychology* (London: Macmillan, 1890), 2:386.

7. Emile Durkheim, *The Elementary Forms of Religious Life* (London: George Allen & Unwin, 1915), conclusion, pp. 415 ff.

104   8. Emile Durkheim, *Suicide* (1897) (London: Routledge &
Kegan Paul, 1952), p. 299.
        9. Emile Durkheim, *Sociology and Philosophy* (1924) (London: Cohen & West, 1953), p. 59.
        10. Jean Piaget, *The Moral Judgment of the Child* (London: Kegan Paul, French & Trubner, 1932).
        11. David Hume, *A Treatise on Human Nature* (1738) (London: J. M. Dent & Sons, 1911), vol. 1, pt. 3, sec. VIII, p. 100.
        12. Henry Sidgwick, *The Methods of Ethics* (1874), 7th ed. (London: Macmillan, 1963), p. 27.
        13. Ibid., p. 28.
        14. Ibid.
        15. Moore, *Principia Ethica*, chap. 1, para. 6.
        16. Harold Himsworth, "The Psychobiology of Human Motivation," in *A Comparison to the Life Sciences,* ed. Stacey B. Day (New York: Van Nostrand, Reinhold & Co., 1979), p. 128.

NINE   SCIENCE AND PHILOSOPHY

        1. Karl Popper, *Conjectures and Refutations,* 2d ed. (London: Routledge & Kegan Paul, 1965), p. 46.
        2. G. E. Moore, *Principia Ethica* (1903) (Cambridge: Cambridge University Press, 1976), p. 54.
        3. Ibid., p. 40.
        4. See Chap. 3 of this volume.
        5. P. B. Medawar, *Induction and Intuition in Scientific Thought* (London: Methuen & Co., 1969), p. 52.
        6. David Hume, *An Inquiry concerning Human Understanding* (1748) (New York: Bobbs-Merrill, 1955), p. 68.
        7. Moore, *Principia Ethica,* pp. 40 and 41.
        8. Ibid., p. 54.
        9. Isaiah Berlin, "Does Political Theory Still Exist?" in *Phi-*

losophy, Politics, and Society, ed. Peter Laslet and W. G.
Runciman (Oxford: Basil Blackwell, 1962), p. 2. See also his
Concepts and Categories (London: Hogarth Press, 1978), p.
144.

Note to Page 97

Index

D

Darwin, Charles: on moral sense and social instinct, 76; and Alfred Russell Wallace, 67–68
Disapprobation, moral, 80
Durkheim, Emile: on group belief and conduct, 79; on suicide, society, and morality, 79

E

Electron microscope, 25
Emotions, relation of, to inherent traits, 75–78, 80
Experience. *See* Understanding, and experience
Experimental method, 96

F

Fact, statements of, 6
Factual experience, as intellectual discipline, 64, 94
Factual observations: accidental, 22–26; generalization of, 29–33; and hypothesis formation, 24–25

G

Gallileo, 25
Generalization, process of, 29–33
Generalizations, descriptive, approximate nature of, 35–37
Generalizations, explanatory: and causal relations, 32–33; degrees of development of, 39; scope of, 58; universality of, 42–43

S

SIR HAROLD HIMSWORTH, as Secretary of the British Medical Research Council from 1949 to 1968, was centrally involved with the development of molecular biology in its formative stages. He was also instrumental in mobilizing the scientific disciplines required to elucidate the medical problems of a nuclear age and in sustaining biomedical research in the tropics during Britain's withdrawal from Empire. Sir Harold came to the council after a distinguished career in research as professor of medicine at the University of London. Now retired, he is a Fellow of the Royal Society, a Fellow of the Royal College of Physicians, and a member of the American Philosophical Society and the American Academy of Arts and Sciences.